Sir Cumference and All the King's Tens

A Math Adventure

Cindy Neuschwander
Illustrated by Wayne Geehan

Charlesbridge

To Don and Maxine Neuschwander, with love—C. N.

To my wonderful grandson, Tomasz Modlinski Geehan, who is at the beginning of his circle of life—W. G.

Published by Charlesbridge
85 Main Street
Watertown, MA 02472
(617) 926-0329
www.charlesbridge.com

Library of Congress Cataloging-in-Publication Data
Neuschwander, Cindy.
Sir Cumference and all the king's tens / Cindy Neuschwander ; illustrated by Wayne Geehan.
p. cm.
Summary: When Sir Cumference and his wife Lady Di of Ameter host a massive surprise birthday party for the king, they must figure out a way to quickly count all the guests who are in attendance.
ISBN 978-1-57091-727-1 (reinforced for library use)
ISBN 978-1-57091-728-8 (softcover)
[1. Counting—Fiction. 2. Parties—Fiction. 3. Kings, queens, rulers, etc.—Fiction. 4. Birthdays—Fiction.] I. Geehan, Wayne, ill. II. Title.
PZ7.N4453Si 2009
[Fic]—dc22 2008025336

Printed in the United States of America
(hc) 10 9 8 7 6 5 4 3 2 1
(sc) 10 9 8 7 6 5 4 3 2 1

Illustrations done in acrylic paint on canvas
Display type and text type set in Opti-Caslon Antique and Dante MT
Color separations by Chroma Graphics, Singapore
Printed and bound by Lake Book Manufacturing, Inc.
Production supervision by Brian G. Walker
Designed by Martha MacLeod Sikkema

"The last time we saw the king, he seemed a bit gloomy," said Lady Di of Ameter to her husband, Sir Cumference, one evening. "His birthday is soon. Let's give him a surprise party here at our castle to cheer him up. We could invite people from the entire countryside."

"That's a fine idea!" answered Sir Cumference. "We'll plan a huge celebration for him."

Lady Di sent out invitations. Servants rushed about the castle, cleaning and cooking. Carpenters built long wooden tables while seamstresses sewed tents. Guests began arriving in groups. Each day more kept coming.

Lady Di showed the guests where to stay. Rooms filled up quickly.

"The castle is already exploding with guests, and an even bigger group is arriving this afternoon," she told Sir Cumference one morning. "King Arthur's party is tonight, and I'm not ready! What a royal mess!"

Sir Cumference nodded. "I'll gather everyone in the meadow to get them out of your way."

Soon a large crowd stood in the grassy field outside the castle walls.

"The king will arrive in a few hours," began Sir Cumference. "Let's practice a Royal March of Greeting. Would everyone step smartly toward the center of the meadow?"

Pandemonium broke out. Knees marching high knocked into arms swinging wide. The Knights of the Round Table crashed into each other, falling into metallic heaps.

Lady Di appeared in the middle of this confusion. "I need to know how many guests will be here for lunch, and then how many for dinner," she called to her husband over the clatter.

Sir Cumference waved his arms. "Attention!" he bellowed. "We need to know how many of you are here." But everyone kept milling around. Counting the crowd seemed like an impossible job.

Sir Kell stepped forward with a suggestion.

"Forming small circles of people might work," he said. "Each group could count its members and call out how many. Lady Di could then add up the numbers."

"Let's try it," said Sir Cumference, shepherding guests into groups.

"61 . . . 111 . . . 58 . . . 17 . . . 46!" cried out voices in the crowd.

"No good," said Lady Di, stopping them. "My head's spinning just trying to keep track of all those numbers."

"We could march by in one straight line," said Sir Lionel Segment, "counting up from one as we pass."

So the group formed a queue. They began moving forward, past Lady Di. The line was so long it disappeared over the hill.

"Too slow," noted Lady Di. "The king's birthday will have come and gone long before I finish figuring this out."

"And I'm getting rather hot standing here!" complained Sir Tangent inside his armor. Others agreed.

"Let's set up some tents," said Sir Cumference. "We can get into the shade while we think of another way to count everyone."

On the edge of the meadow, the castle workers erected a small tent. Immediately knights, ladies, and villagers rushed inside. It bulged dangerously.

"That tent is just too tiny," huffed Sir Tangent as he walked out. "It doesn't even hold ten people." He threw up his hands in impatience.

"Maybe this party for the king was a bad idea," Lady Di said to Sir Cumference. "Tempers are beginning to flare, and I still don't know how many meals to serve."

“We need a new idea,” said Sir Cumference. Then he paused and smiled. “Or maybe we can use parts of everyone’s ideas. I think I have the solution!”

He stepped into the middle of the meadow.

"Attention, everyone! Please gather into small groups, as Sir Kell suggested. Spread those groups out into lines, as Sir Lionel Segment described. Each line should have ten people, like the ten fingers on Sir Tangent's hands."

The crowd grouped themselves as Sir Cumference directed. Lady Di started to count, but there were still so many rows!

"If we put ten rows together, they would equal one hundred," she said. "That would make this counting go even more quickly." The hot but patient guests moved into the larger formations.

"I'm counting nine groups of one hundred," said Lady Di. "There are also eight rows of ten and one row with only seven. That's 900 plus 80 plus 7. Now at least I know how many lunches we need."

Then twenty-five more people arrived from the small town of Lower Numberton.

"Welcome!" said Lady Di, smiling. "Could you get together in rows of ten, and could three of you join that line of seven to make another row of ten?"

Two new groups of ten joined the other eight rows of ten and made a new group of one hundred. One more row of ten remained, with a lone farmer and his wife standing shyly just beside it.

"Oh my! Now we're up to 1,012 guests," murmured Lady Di.

Several more tents of different sizes were set up to provide shade for everyone. To the left of the tiny tent that fit nine people or fewer, the castle workers pitched a bigger tent. It could hold up to nine rows of ten, or ninety. Next to that tent there was an even bigger one for as many as nine groups of one hundred. An enormous tent was next, for crowds of up to nine thousand.

"We'll serve lunch in the largest tent," declared Lady Di.

As everyone was finishing the last luncheon bites, a cloud of dust appeared in the distance. More guests were coming! It was a huge group from the king's city, Camelot.

"Greetings, Sir Cumference!" called the leader of the caravan. "We're here for the party. The king and his nobles should arrive shortly."

"Well then," said Sir Cumference, rubbing his hands together briskly, "let's find a place for all of you."

It took a while, but the Camelot guests were finally organized. They had eight groups of one thousand, nine groups of one hundred, eight groups of ten, and seven singles. Lady Di sent a messenger back to the castle to add 8,987 to the 1,012 guests who were already there for the evening feast.

"We'll need to put up another huge tent for dinner," said Sir Cumference.

"Yes," agreed Lady Di. "And we could really surprise the king if everyone stayed hidden until his arrival."

So they assigned each guest to a tent. The meadow became a Knightly Number Neighborhood. There were nine folks in the first tent. There were ninety in the second tent next door. There were nine hundred in the third tent, and nine thousand in the fourth tent. The fifth tent was empty except for long wooden tables and benches.

Lady Di passed out silk scarves to each person, saying, "The color of your scarf matches your tent flag."

Then the entire group gathered to practice their surprise greeting for the king. They had rehearsed only once when a trumpet blared in the distance. *Bap, ba-ba BAH!*

"The king is arriving!" yelled Sir Cumference. "Hurry into your tents and wait for my signal!"

When the king arrived, he had a sour look on his face.

"It's my birthday," he muttered grumpily, "and I've been stuck on the back of a horse for hours! What are all these tents for? I'm in no mood for a jousting tournament."

Sir Cumference and Lady Di welcomed the king.

"Your Majesty," said Lady Di, showing him to his seat of honor.

"A birthday greeting for you," said Sir Cumference, bowing and clapping his hands. This was the signal the hidden guests were waiting for.

They streamed out of their tents, singing and dancing. The king was surprised. His sour look began to turn into a smile.

At the end of their performance, everyone lifted and lowered their scarves one row after another, starting with the largest group of nine thousand, then the nine hundred, then the ninety, and finally the nine.

It looked as if a shimmering ocean wave had crashed at the feet of the king.

He stood and applauded. "Well done! Thank you, my friends!"

Lady Di clapped her hands. "And now, let's eat!"

After a tasty dinner, the cooks brought out an enormous birthday cake and other delicious sweets. The guests cheered as the king blew out the candles.

Then everyone heard the thunder of hoofbeats. "Uh-oh!" said Sir Cumference. "Another group approaches. Their flag shows them to be from the city of Addingmoor."

"How many more tents will we need?" he wondered out loud.

"And how many more desserts will that be?" asked Lady Di. "I hope we have enough cake!"

The concept of place value, or where numbers "live," allows us to make any number using the digits 0–9. A digit's place tells its value, or how much it is worth.

In the story 9,999 guests show up for King Arthur's party. The number tells us there are nine groups of one thousand, nine groups of one hundred, nine groups of ten, and nine single guests. The guests fill the tents in a Knightly Number Neighborhood.

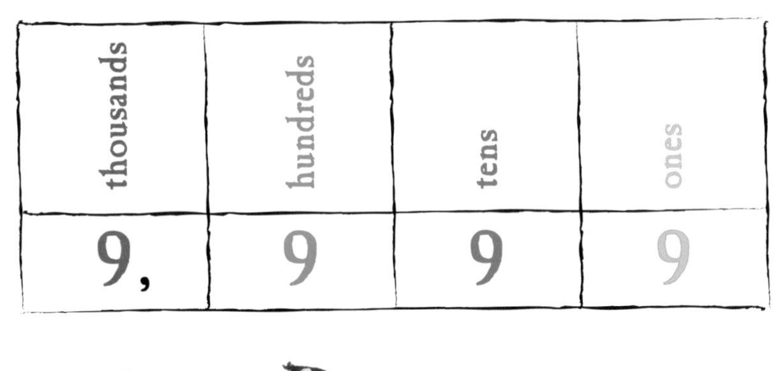

thousands	hundreds	tens	ones
9,	9	9	9

The next time you see a big number, try imagining that the digits live in a Number Neighborhood. Each digit's place, or "tent," is ten times bigger than the one to its right.